Access to Space in the Southern Hemisphere

Access to Space in the Southern Hemisphere

Final Report

Adelaide
2020

Agathon: Access to Space in the Southern Hemisphere
Volume 7, 2020

Agathon refers to the Greek word used by Plato in the *Republic* to refer to 'the good beyond being', most notably deployed in recent times by Iris Murdoch and Emmanuel Levinas, both of whom use this touchstone to situate ethics at the heart of all of philosophy

We live in an evolving and increasingly complex world but our ethical concepts have frequently struggled to keep pace with the change. As a result, much of what passes for public debate at present remains in the grip of either deterministic or consequentialist thinking, both built on outdated assumptions and both representing attempts to address major issues in the absence of ethical concepts. We suffer, as Iris Murdoch lamented, a 'loss of concepts, the loss of a moral and political vocabulary'.

The interdisciplinary journal, *agathon*, seeks to bring together scholars from across the humanities, social sciences and sciences, including disciplines such as philosophy, theology, law and medicine, to engage with the ethical questions that now beset the modern world.

The journal is a home for considering questions such as how we deal with competing values in ethical discourse, how ethical theory finds expression in practice, what constitutes ethical character and how it is cultivated, and what excellence and wisdom look like for the ethical person or society in the twenty-first century.

Agathon is an international and interdisciplinary refereed journal published annually by ATF Press.

Chief Editor
Dr Paul Babie, University of Adelaide Law School Professor of the Theory and Law of Property

Editorial Board
- Professor Terence Lovat, Editor in Chief Bonhoeffer Legacy, The University of Newcastle, Australia & Hon Fellow University of Oxford, UK.
- Professor Robert Crotty, former Director, Ethics Centre of South Australia, Emeritus Professor of Religion and Education, University of South Australia, Adelaide.
- Managing Editor and Publisher Mr Hilary Regan, Publisher, ATF Press Publishing Group, PO Box 234 Brompton, SA 5007, Australia. Email: hdregan@atf.org.au

Subscription Rates
Local: Individual Aus $55 Institutions Aus $65 Overseas: Individuals US $60 Institutions US $65

Agathon is published by ATF Press an imprint of the ATF Press Publishing Group which is owned by ATF (Australia) Ltd (ABN 90 116 359 963) and is published once a year. ISSN 2201-3563

The 2020 Southern Hemisphere Space Studies Program was held at the Mawson Lakes campus of the University of South Australia (UniSA), Adelaide, by the International Space University (ISU) and UniSA.

ISBN: 978-1-925371-66-6 soft
         978-1-925371-67-3 hardback
         978-1-925371-68-0 epub
         978-1-925371-69-7 pdf

Published by:

An imprint of the ATF Press Publishing
Group owned by ATF (Australia) Ltd.
PO Box 234
Brompton, SA 5007
Australia
ABN 90 116 359 963
www.atfpress.com
Making a lasting impact

Table of Contents

Participant List

Anu Rajendran

Brianna Ganly

Joshua Kahn

Rebekah Holliday

Samuel Hobbs

Steven Despotellis

Zvonko Vidos

Guy Marcus

Christian Thaler-Wolski

Hongxing Zuo

Kaimin Zhou

Na Zhang

Qiwei Guo

Shideng Yang

Shiyin Lyu

Si Li

Xiangyu (Sean) Li

Yiming Zhou

Yonggang Luo

Andrea Perin

Luca Santachiara

Preeti

Sujith Rajan

Sasiluck Thammasit

Ibnu Rusydi

Fatima AlShamsi

The authors gratefully acknowledge the generous guidance, support, and direction provided by the following faculty, visiting lecturers, teaching associates, program staff, advisors, and experts:

Ms Manal Al-Rasheed, **University of Central Florida**

Mr Hugo André Costa, **LSE Space GmbH**

Mr Adrià Argemí Samsó, **Pangea Aerospace**

Dr Jacques Arnould, **CNES - Headquarters**

Mr Sebastien Bessat, **International Space University (ISU)**

Ms Philomena Bonis, **Queensmount Senior Public School**

Assoc Prof David Bruce, **University of South Australia (UniSA)**

Prof John Connolly, **NASA Johnson Space Center**

Mr David Cowdrey, **University of South Australia (UniSA)**

Mr Paul Curnow, **University of South Australia (UniSA)**

Mr Lloyd Damp, **Southern Launch**

Dr Malcolm Davis, **Australian Strategic Policy Institute**

Mr Michael Davis, **Space Industry Association of Australia**

President Juan de Dalmau, **International Space University (ISU)**

Prof Kerrie Dougherty, **University of New South Wales**

Prof Alan Duffy, **Swinburne University of Technology**

Prof Steven Freeland, **University of Western Sydney**

Mr Ryo Futamata, **Nihon University**

Mr Alessandro Gabrielli, **Avio S.p.A.**

Ms Michelle Gilmour, **Gilmour Space Technologies**

Dr Brett Gooden, **University of South Australia (UniSA)**

Assoc Prof Alice Gorman, **Flinders University**

Mr Thomas Goulding, **Royal Melbourne Institute of Technology**

Dr James Green, **NASA Headquarters**

Mr Daniel Griffiths, **University of South Australia (UniSA)**

Mr Rainer Horn, **SpaceTec Partners GmbH**

Mr Femi Ishola, **Kyushu Institute of Technology**

Assoc Prof Ady James, **University of South Australia (UniSA)**

Mr Mark Jessop, **Amateur Radio Experimenters Group**

Ms Teneille Johnson, **University of South Australia (UniSA)**

Ms Amanda Johnston, **University of South Australia (UniSA)**

Mr Goktug Karacalioglu, **International Space University (ISU)**

Dr Justin Karl, **National Aeronautics and Space Administration**

Ms Rei Kawashima, **HEPTA-Sat**

Ms Mina Konaka, **Tohoku University**

Principal Donna Lawler, **Azimuth Advisory Pty Ltd**

Mr Eamon Lawson, **University of South Australia (UniSA)**

Prof Gottfried Lechner, **University of South Australia (UniSA)**

Mr Martin Lewicki, **University of South Australia (UniSA)**

Mr Anderson Liew, **HSBC**

Mr Tony Lindsay, **Lockheed Martin**

Dr Charley Lineweaver, **Australian National University**

Mr Darin Lovett, **South Australian Space Industry Center**

Mr Gary Martin, **International Space University (ISU)**

Ms Ruth McAvinia, **ATG Europe**

Dr Katarina Miljkovic, **Curtin University**

Dr Val Munsani, **South African National Space Agency**

Dr Paolo Nespoli, **NASA Johnson Space Center**

Dr David Neudegg, **Bureau of Meteorology**

Dr Patrick Neumann, **Neumann Space**

Dr Kimberley Norris, **University of Tasmania**

Dr Joseph O'Leary, **Electro Optic Systems Pty Limited**

Prof Walter Peeters, **International Space University (ISU)**

Ms Lindsey Pollock, **University of South Australia (UniSA)**

Mr Scott Pollock, **Southern Cross Pyrotechnics**

Mr Andrea Preve, **Avio S.p.A.**

Ms Anisha Rajmane, **KIT College of Engineering**

Ms Alexandra Ryan, **International Space University (ISU)**

Mr Scott Schneider, **Leiden University**

Dr Jan Walter Schroeder, **CisLunar Industries**

Mr Tim Searle, **New Zealand Space Agency**

Mr Michael Siddall, **University of South Australia (UniSA)**

Mr Noel Siemon, **International Space University (ISU)**

Dr Michael Simpson, **Secure World Foundation**

Prof Alan Smith, **University College London**

Prof Parwati Sofan, **University of South Australia (UniSA)**

Dr Su-Yin Tan, **University of Waterloo**

Mr Robertus Triharjanto, **Indonesian National Institute of**

Aeronautics and Space

Mr Scott Wallis, **Equatorial Launch Australia**

Ms Hannah Webber, **University of South Australia (UniSA)**

Asst Prof Masahiko Yamazaki, **Nihon University**

Mr Emil Zankov, **University of South Australia (UniSA)**

Mr Taiga Zengo, **Nihon University**

Mr Alexander Ziegele, **DSTG**

Program Sponsors

ATF Press
Lockheed Martin
National Aeronautics and Space Administration
Portugal Space Agency
Sir Ross and Sir Keith Smith Fund
Taylors Wines
Tenth to the Ninth plus Foundation
The Aerospace Corporation
The Simeone Group

Sponsored Placements

Australian Space Agency
China Aerospace Science and Technology Corporation
China Satellite Launch and Tracking Control General
Commonwealth Scientific and Industrial Research Organisation
Indian Space Research Organization
National Aeronautics and Space Administration
Nova Systems
United Arab Emirates Space Agency

Scholarship Providers

Asia Pacific Satellite Communications Council
European Space Agency
Italian Space Agency
Sir Ross and Sir Keith Smith Fund
South Australian Space Industry Centre
(Government of South Australia)

Event Sponsors

Amateur Radio Experimenters Group
City of Salisbury
Cleland Wildlife Park (SA Department for Environment and Water)
Space Industry Association of Australia
Vex Robotics

Faculty, staff, visiting lecturers and experts who were consulted for the Team Project.

Program Sponsors

ATF Press
Aerospace Corporation
Lockheed Martin
National Aeronautics and Space Administration
Portugal Space Agency
Sir Ross and Sir Keith Smith Fund
Taylors Wines
Tenth to the Ninth plus Foundation

Organization-Sponsored Places

Australian Department of Defence
Australian Space Agency
China Aerospace Science and Technology Corporation
China Satellite Launch and Tracking Control General
Commonwealth Scientific and Industrial Research Organisation
Indian Space Research Organization
National Aeronautics and Space Administration
Nova Systems
United Arab Emirates Space Agency

Event Sponsors

Amateur Radio Experimenters Group
City of Salisbury
Cleland Wildlife Park (SA Department for Environment and Water)
Space Industry Association of Australia
Vex Robotics

Scholarship Providers

Asia Pacific Satellite Communications Council
European Space Agency
Italian Space Agency
Sir Ross and Sir Keith Smith Fund
South Australian Space Industry Centre (Government of South Australia)

The Shssp20 Mission Patch

The SHSSP20 mission patch depicts a rocket that is taking off from the Moon to go to Mars. This is our way to acknowledge Project Artemis and humankind's reinvigorated willingness to venture into new territories with human spaceflight. We honour the wildlife in Australia that has recently suffered in devastating bushfires. We wish the koala to be understood as a symbol for all of humankind that may eventually find refuge on other planets.

Abstract

Access To Space

The space sector worldwide is experiencing a shift from traditional activities involving large heavy launch vehicles and satellites to a much faster paced NewSpace paradigm. NewSpace is primarily driven by private organizations and is characterised by smaller and lighter missions involving Smallsats. The space launch sector is primarily the domain of the Northern Hemisphere, however, this is beginning to change as Southern Hemisphere nations seek to pursue access to space in their own right. This report considers three scenarios through which Southern Hemisphere nations may pursue access to space. The third of these scenarios is then expanded upon to explore the benefits of an international collaborative framework to facilitate this access. Australia is used as a case study to identify considerations for enabling access within the framework.

Faculty Preface

ISU UNISA

The International Space University (ISU) has maintained a unique leadership position within the space community for more than 30 years.

Since 2011, ISU has held the Southern Hemisphere Space Studies Program (SHSSP) partnered with the University of South Australia (UniSA). Our colleagues at UniSA throughout this time have been welcoming, gracious, and very helpful. We thank the program director Göktuğ Karacalıoğlu and co-director Ady James, as well as logistic coordinator Amanda Johnston for their dedicated work to make this SHSSP 2020 session a grand success. We acknowledge that we met on the traditional lands for the Kaurna people for the duration of this program.

An important part of the SHSSP is the Team Project (TP) in which participants work on discussing interesting questions and possible solutions in the space community. In presenting ideas for the future of sustainable space activities, the participants undertake the five-week program in an international, intercultural, and interdisciplinary group. This process helps them develop the essential skills complementing their career in the global space sector.

This report is the result of the SHSSP 2020 TP on "Access to Space". Participants were tasked to investigate the different options and opportunities of what access to space means for the Southern Hemisphere, choosing Australia as a case study. This was an especially challenging project due to the broad scope and the many interrelated issues posed by the problem. They had to accomplish this difficult job in the midst of a busy schedule of other program commitments and with tight deadlines.

In this 9th anniversary year of the SHSSP, we are proud of what this talented team of participants has accomplished in such a short time. We look forward to seeing how each member applies their skills beyond the program, and we look forward to working with them as friends, colleagues, and members of the space family.

Dr. Jan Walter Schroeder TP Chair
Scott Schneider TP Associate Chair
Eamon Lawson Teaching Associate
Mina Konaka Teaching Associate

Participant Preface

TP UNISA

Space has inspired humans on Earth and provided tools for survival since time immemorial. For tens of thousands of years following ancient time, Aboriginal and Torres Strait Islander peoples (in the country known today as Australia) have used the stars for navigation, timekeeping, education and storytelling as well as to ensure sustainability of natural resources (Norris 2014).

When we look to the night sky today from Australia in this age of commercialized space exploration, as this report's project team has done over the past five weeks, the sky we (can still) see may look different yet is somewhat similarly utilized. In fact, we still use objects in the sky to assist us with life on Earth.

From timekeeping, communications, and navigation to natural resource and emergency management, data provided by satellites in space are almost completely interwoven in how we live. We are inspired by the knowledge that fellow humans are passing overhead on the International Space Station. We draw inspiration from plans for future exploration, science, and tourism opportunities in space and how these activities could improve and sustain life on Earth.

With this in mind, it is more important than ever for decision makers to pursue future space exploration and potential colonization of planets in the same spirit our ancestors did when the space age began – peacefully. Lessons learned on Earth can guide our efforts to care for the space environment we are so reliant on in a unified way.

It is hoped our report can present decision makers with a case for peaceful and sustainable space-related activities at home and far away, wherever the journey takes us.

SHSSP2020
Team: Access to Space

Acronyms

AEB	Brazilian Space Agency
ASEAN	Association of South Eastern Asian Nations
AVIO	AVIO S.p.A (Italian Aerospace company)
COTS	Commercial-Off-The-Shelf
ELA	Equatorial Launch Australia
EO	Earth Observation
ESA	European Space Agency
FAA	Federal Aviation Authority
GCRI	Global Climate Risk Index
GDP	Gross Domestic Product
GNSS	Global Navigation Satellite Systems
ICBM	Intercontinental Ballistic Missile
ISRO	Indian Space Research Organisation
ISU	International Space University
JPL	Jet Propulsion Laboratory
LAPAN	Indonesia's National Institute of Aeronautics and Space
LEO	Low Earth Orbit
MEO	Medium Earth Orbit
NASA	National Aeronautics and Space Administration
NT	Northern Territory (Australia)
NZ	New Zealand/Aotearoa
PE	Private Equity

PPP	Public-Private-Partnership
R&D	Research and Development
SAR	Synthetic Aperture Radar
SATCOM	Satellite Communications
SL	Southern Launch
SSO	Sun Synchronous Orbit
SWOT	Strengths, Weaknesses, Opportunities, and Threats
UK	United Kingdom
UN	United Nations
USA	United States of America
VC	Venture Capital

List of Figures

List of Tables

Terminology

Access to Space

Traditionally, access to space is equated with having the ability to physically visit space through the use of a launch vehicle and, if desired, place an object in orbit. However, there are multiple ways in which access to space may be defined: launch capability, access to data from sovereign satellites or access to data from third party satellites as shown in Figure 1. Countries that do not have capacity to build and launch their own rockets still need to access space for military, civil and economic factors. To date, for example, Australia has facilitated its needs by focusing on sharing the in-space capabilities of commercial companies or international partners who own and operate the satellites (Australian Government, 2013 and Clark et. al., 2018).

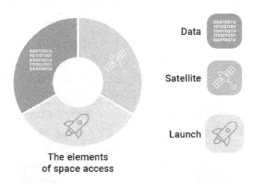

Figure 1: Ways in which access to space could be fulfilled

Smallsats

The definition of Smallsats varies between market research organisations worldwide. Market research-established categories are Pico/Femto satellites (<1kg), Nano satellites (<10 kg), Micro satellites (10-100kg) and Mini satellites (100-500kg) which are collectively known as Smallsats (Euroconsult, 2016 and Bryce, 2020). Bryce Space and Technology considers the same categories but defines Micro and Mini satellites as 11-200kg and 201-600kg respectively. This report considers Smallsats as satellites under 600kg.

Southern Hemisphere

This report considers a nation to be in the Southern Hemisphere if a significant part of its landmass is located South of the equator.

Newspace

Space, like other areas of technology, is experiencing the effects of miniaturisation of components, fast, iterative software development, and new additive manufacturing technologies. Additionally, in the NewSpace era private financing and entrepreneurship in both the upstream and downstream space domains drive commercial adoption. This report considers the combination of these factors to constitute NewSpace.

Launcher, Launch Vehicle, Spaceports, Launch Site Facility, Launch (Service) Provider

When referring to the vehicle that is launched to place payloads in orbit, the project team prefers the term launch vehicle over other terms such as launcher. In the past, launch vehicles were almost always coupled with specific launch sites. With the exception of Rocket Lab and SpaceX, in NewSpace we will see more cases of a separation between launch vehicle and launch sites because of more flexibility and "plug and play"- modularity between launch vehicles and sites.

The term spaceport is used to describe large complexes similar to airports. Spaceports are used for human space flight such as Virgin Galactic's suborbital experience. Launch site facility denotes the character of commercial business-to-business operations with access for authorized personnel only.

Introduction

Mission Statement: To explore and assess the implementation of NewSpace access for the Southern Hemisphere by evaluating strategic, economic, political and environmental factors to inform government and commercial decision making.

> "NewSpace refers to the recent commercialisation of the space sector. While the state used to have a monopoly over the sector, private actors now play an increasingly important role . . ."
>
> –Airbus (2020)

An emerging trend within the global space sector throughout the last decade is the increasing focus on smaller, cheaper, and faster space technology development which has been called the NewSpace revolution (Datta, 2017). In the 20th Century, space was solely the province of States, such as the United States of America (USA) and Russia, capable of funding complex and expensive space projects. However, a paradigm shift is occurring based on smaller, cheaper electronics and advanced manufacturing techniques, which is acting as an enabler for smaller States to develop access to space capability. The NewSpace paradigm is driven by the establishment of companies backed by private investment and entrepreneurship; the emergence of Rocket Lab in New Zealand is one example.

Despite the increasing globalisation of the space industry, all but three of the launch sites active today are located in the Northern Hemisphere (Gunter, 2020). However, we are beginning to see increased urgency and awareness being placed on access to space by Southern Hemisphere nations. In addition to New Zealand, Australia,

Indonesia, and Brazil are beginning to actively pursue space access through NewSpace companies (Boadle, 2018 and Department for Trade and Investment, 2020).

Even with the increased activity in the Southern Hemisphere, it remains to be seen how access to space for Southern Hemisphere nations will be achieved in the NewSpace era. Exploration of how this access may be facilitated is the focus of this report. The report takes the perspective of what is beneficial for the Southern Hemisphere as a whole rather than specific States.

Methodology

This Report Aims to:

- Evaluate the current and future demand for Smallsats and associated launches.
- Explore how space access may be facilitated within the Southern Hemisphere.
- Analyze the specific role and implications for Australia in implementing space access.
- Provide considerations for government and industry policy makers.

This report is being submitted to the International Space University (ISU) to fulfil the requirements of the Southern Hemisphere Space Studies Program (SHSSP). This report is one of three deliverables for the team research project, with the other two being a corresponding executive summary and an academic presentation. The SHSSP was five weeks in duration with two weeks dedicated to the team research project.

The project team for this report consisted of 26 people with diverse backgrounds from nine countries. The team was structured with a management committee of six people, two of which were Lead Project Managers who supervised the project. They were assisted by four co-project managers who supervized the content of deliverables. The remaining team members were divided into four research teams each focusing on either economics, government, environment or strategic considerations of "Access to Space from the Southern Hemisphere".

The project team has equated access to space with the ability to launch and place something in orbit. This does not necessarily mean that a nation has to have a sovereign launch capability. In the context of this report it means that they must have a reliable, efficient and effective means of placing a commercial satellite in orbit, even if through a partner country.

Each of the four research teams conducted their own literature review. In total, 19 external experts from industry and government were interviewed either in person, via video link, over the phone or via email. This research contributed to identifying three different scenarios for access to space in the Southern Hemisphere which were then assessed using a Strength, Weaknesses, Opportunities and Threats (SWOT) analysis. A SWOT analysis survey was completed by members of the ISU community associated with the SHSSP20 program. The team used key metrics identified in the mission statement for a comparative analysis of the SWOT findings to determine the scenario to then be applied to Australia as a case study.

Market Analysis

Over the last few years the number of Smallsats has increased.

In Figure 2, the launch of constellations with satellites with mass less than 51kg is the main market driver. In terms of mission category, 80 percent of 250-500kg satellites are to support technology development, science and government Earth Observation (EO) applications. In contrast, satellites of 250kg and below contribute to satellite communications (SATCOM) and EO applications.

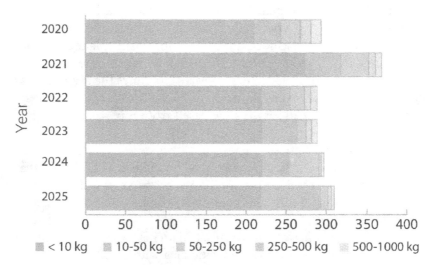

Figure 2: Projected SmallSats Launched by Mass Category (Source: Modified from Prospects for the Small Satellite Market; EuroConsult Executive report July 2016 - Page 18)

The purpose of this section is to describe and analyze the global market situation for Smallsats, launch vehicles and launch sites. The project team considers these three interrelated components the most important aspects of access to space. Launch vehicles and launch facilities are fundamentally linked and both require customers who wish to transport a payload to orbit. Launch services are normally provided by either the launch site operator or the launch vehicle manufacturer or, in the case of brokers, by an independent third party.

Over the last few years the number of Smallsats launched has increased. The market course corrected in 2019 and is now projected to grow in a stable manner (Bryce, 2019).

The Federal Aviation Authority (FAA) predicted in 2018 that from 2019–2029 2864 payloads will be launched (FAA, 2018). Mini and Nano satellites often have a lifespan of only two to three years due to low orbits with high atmospheric drag (Masunaga, 2018). Satellites in the 100kg category often have a lifespan in the range of five to seven years. If the FAA prediction is evenly distributed and an average three year lifespan with a two-thirds replenishment rate is applied, a base forecast of 324 Smallsats per year can be assumed. Italian large launch vehicle manufacturer Avio S.p.a (manufacturer of the Ariane and VEGA family of rockets), together with Euroconsult, forecasts 5,111 satellites in the mass category up to 200kg will be launched between 2020–2030 (Avio, 2020). According to Bryce (2020) the 2019 SmallSat highlights included an increase in SmallSat launches from 24 percent in 2012 to 45 percent in 2019 of all satellite launches with an average mass of 109 kg. That is nearly two times the average mass from 2018, and six times the mass of 2017. The share of SmallSats providing commercial services increased from 6 percent in 2012 to 62 percent in 2019.

The difference between Bryce and Euroconsult may result from taking into account large commercial satellite constellations that are planned to come online in the next few years. Alternatively, these forecasts may differ because of different assumptions on said replenishment rate.

2019 saw an increase in Smallsat launches from 24 percent in 2012 to 45 percent of all satellite launches with an average mass of 109kg which is nearly two times the average mass from 2018, and six times the mass of 2017. The share of SmallSats providing commercial services increased from six percent in 2012 to 62 percent in 2019 (Bryce Space and Technology, 2020).

SmallSats are used for:

- Remote sensing (earth observation). Examples include Lemur-2, Spire's ship and aircraft-tracking satellite constellation;
- Communications. For example, the Keppler Communications KIPP satellite which provides backhaul for Internet of Things (IoT) and Machine-to-Machine (M2M) communications is an example;
- Science satellites. An example is NASA's Jet Propulsion Laboratory's (JPL) MarCO-A/Wall-E, which provides real communication mission support for National Aeronautics and Space Administration (NASA) InSight Mars Lander;
- Technology satellites. For example, Tongchuan which demonstrated space-to-space inter-satellite and space-to-Earth communications, and
- Completely novel applications such as Enoch, a passive CubeSat by the LA County Museum of Art (DelPozzo et. al., 2018).

Commercial satellite constellations are a driver of Smallsat demand because they often require a large number of Smallsats operating at one time. Table 1 provides examples of commercial Smallsat constellations. The composition of Smallsats by area of activity varies yearly (refer Figure 3); the variation is a function of the different types of commercial constellations launched. There were 328 Smallsats launched in 2018. Of these approximately 37 percent performed remote sensing functions, 40 percent for technology development and 11 percent in communication. In contrast, 389 Smallsats were launched in 2019–37 percent in the communication category, 26 percent were in remote sensing, and 32 percent in technology development. Of particular note, in both years less than three percent of launched SmallSats were for military or intelligence use, making the vast majority designated for commercial and civil use.

(Bryce Space and Technology, 2019 and Bryce Space and Technology, 2020).

Table 1: Examples of Existing or Planned Satellite Constellations. Sources (Henry, 2019), (Nanosats Database, 2020), (Escher, 2018), (Myriota, 2019) and (Bbc News, 2018).

Constellations	Category	Usage Classification	Comments
Starlink	Mega	Communications	Currently planned 12,000 satellites; could go up to 42,000
Spire	Large	(AIS) vessel tracking	Launched 115, 150 planned.
Planet	Large	Earth Observation (EO)	Launched 387 sats in 4 years, 150 are active.
Fleet	Medium	IoT / M2M Communication	4 active sats, 100 planned.
Myriota	Medium	IoT / M2M Communication	1 test satellite launched; partnering with manufacturers
Iceye	Medium	SAR Remote Sensing	18 sats operational

There are two models for Smallsats launches that are characteristics of the NewSpace industry:

1. Rideshare with large launch vehicles that deploy a primary payload into orbit and then a number of rideshare nano satellites at the same time (sometimes up to 20 or 30). This model is usually cost-effective as the fixed costs of the launch have largely been covered by the primary payload. A drawback is that smaller payload customers need to adjust their schedule to suit the main payload customer. Sometimes, they may get transferred to other launches. An improved model of the above is a "taxi service" as proposed by SpaceX, where the launch operator schedules regular launches and markets the capacity (Etherington, 2019). In this model the launch occurs regardless of whether all capacity has been sold or not. An additional advantage is that SpaceX then releases the payload into the desired orbit like a taxi dropping off passengers.

2. Dedicated launches that carry a single Smallsat into space are economically feasible only for larger Smallsats greater than 50kg but the costs of Smallsat design and manufacturing in the micro and nano satellite range is coming down to a few hundred thousand dollars (Masunaga, 2018). Prospected costs of dedicated medium and mini launch vehicles need to be in line with the total cost of deploying a satellite.

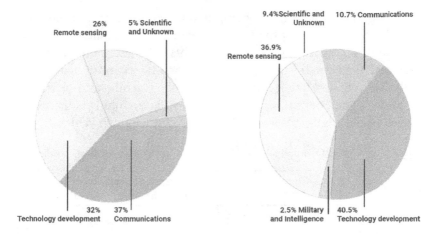

Figure 3: Launches In 2018 and 2019 by Activity (Source: Smallsats by the Numbers 2019) and Smallsats by the Numbers 2020 (Bryce Space and Technology, 2019 (Page 6) and Bryce Space and Technology, 2020 (Page 10)).

To date, there are 34 orbital launch facilities in 17 countries (Gunter, 2020), however, not all are available for commercial launch as some launch facilities are restricted to military usage or in embargoed countries.

The launch facilities with the greatest launch percentage for commercial Low Earth Orbit (LEO) Smallsat payloads are in USA, Russia, New Zealand, China, France, and India (Seradata, 2020).

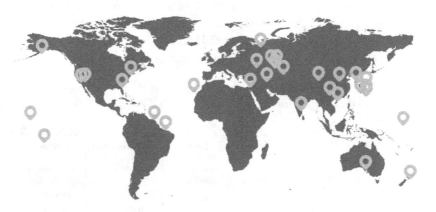

Figure 4: Worldwide Launch Facilities (Orbital) Source: Gunter, 2020.

New launch facilities are being developed for mini and medium launch vehicles to meet the increased demand for Smallsat payloads in either rideshare or dedicated flights. For example, Portugal (Azores), the United Kingdom (Melness, Scotland), Indonesia (Biak), Italy (Taranto-Grottaglie), Australia (South Australia and Northern Territory), and Spain (Canary Islands) are all developing launch sites targeting the combination of mini and medium launch vehicles.

Table 2: The Top 5 Operational Small Sat Launch Sites Worldwide for Leo Orbit

Launch Sites/ Countries	Accumulated percentage share of Launches for LEO Smallsats
USA: Cape Canaveral and Vandenberg Air Force Base	39%
Russia: Yasny and Baikonur	25%
Jiuquan Satellite Launch Center, China	8%
Guiana, France	8%
Shriharikota, India	5%
Other	15%

In the Southern Hemisphere, New Zealand's Rocket Lab has proven capability with 11 successful launches from their site on the Mahia Peninsula and is developing a second launchpad to increase capacity (Foust, 2019), (Spence, 2019), (Waters, 2019), (The Jakarta Post, 2019) and (Moon, 2019).

In response to the increased commercial demand in Smallsat launches, more than 50 dedicated Smallsat launch vehicles are in development (Bryce Space and Technology, 2019). Launches dedicated to Smallsats have increased from approximately 10 launches to 27 between 2017 and 2019 (Bryce Space and Technology, 2020).

Table 3: Notable Future Launch Sites in or Near The Southern Hemisphere.
Sources: (Foust, 2019), (Spence, 2019), (Waters, 2019), (The Jakarta Post, 2019) and (Moon, 2019).

Country	Name/ Location	Latitude	Type	Tested launch vehicle	Comments
Guam (US)	Andersen AFB	13.4° North	Equatorial	Virgin Galactic, Launcher One	announced for 2020
Indonesia	Biak	1° South	Equatorial		announced for 2024

New Zealand	Mahai Launch Site, launch pad 2	39.1° South	Polar and SSO	Electron	announced for 2020
Australia	Whalers Way, South Australia	34.9° South	Polar and SSO	Perigee	announced for 2020
Australia	Arnhem Space Centre, NT	12.1° South	Equatorial		announced for 2020

Current capability to launch Smallsats has existed primarily with spacecraft that offer rideshare services, such as SpaceX with the Falcon 9, Russia with the Soyuz-2, and now dedicated Smallsat launch vehicles, such as Rocket Lab's Electron (Gunter, 2020). Smallsat launch vehicles are being designed with novel and unique technologies in mind in order to reduce cost of launch and thus price-per-kg; companies with larger rockets, such as SpaceX, have enough mass capability to offer prices much lower than current dedicated Smallsat launch vehicles.

However, rideshare launch vehicles are less convenient in placing Smallsats in dedicated orbits given that the rideshare payloads are often not the main payload. Therefore, as numbers of Smallsats increase and the need for replenishment in dedicated orbits increases, it follows that the number of dedicated launches and dedicated launch vehicles will increase.

Some Smallsat launch vehicles rely on portable infrastructure methods to lower cost, such as those manufactured by ABL Space Systems. Other companies use new propellants or propellant technology, such as Orbex (using propane and liquid oxygen as a bipropellant) or Gilmour Space Technologies (with hybrid rocket motors). There are also completely novel developments of technology that could replace current launch mechanism, such as Spinlaunch which would use mass accelerator technology to deliver payloads into orbit.

The following table includes a selection of notable launch vehicles that are currently in development. Rocket Lab's Electron is mentioned for context (raised US$215m funding) and Spinlaunch is listed because it is a novel and unique idea (Crunchbase, 2020).

Table 4: Selection of Launch Vehicles in Development. Source: Crunchbase, 2020.

Name	Country of Origin	US$ Funding raised	Comments
Spinlaunch	United States	$80M	Catapult or centrifuge to spin a rocket
Perigee	South Korea	$12M	Launching at Whalers Way, SA, in 2020.
ABL Space Systems	United States	N/A	Transportable system
Gilmour Space Technologies	Australia	$24M	Launching in 2021
Orbex	United Kingdom	$40M	Launching in 2021
Vega Light	Italy	N/A	Launching in 2022
RocketLab	United States/New Zealand	$215M	First Smallsat launcher, started in US, "returned to NZ"

In Summary

The Smallsat market is varied and growing, especially in the nano satellite market. In particular, Smallsats serving the communications industry will lead to the largest deployment of satellites currently in orbit, with SpaceX's Starlink leading the way for many companies to follow. In conjunction with the increasing number of Smallsats is the increasing development and deployment of dedicated SmallSat launch vehicles across many countries.

Indian Space Research Organization (ISRO) and SpaceX are currently launching the majority of these Smallsats at prices lower than other launch providers (SpaceDaily, 2018); however, as the number of Smallsats increase, the number of Smallsats requiring specific orbits will also increase, leading to the usage of more dedicated launches over rideshare services. Combined with the availability of low-cost mini-launch vehicles, this could drive down the cost of these services.

With these new developments, access to space is now more accessible than ever before.

Need for Access

The market survey in the previous section suggests that the demand for Smallsats and NewSpace launches is likely to increase. Even with NewSpace being driven primarily by private investment, State support for space access is required to facilitate the regulation and growth of these companies. This report is concerned with the implementation of space access in the Southern Hemisphere and as such two key questions are considered:

1. Why would a State pursue space access capabilities and how can NewSpace provide an effective and efficient avenue?
2. What are the benefits of launching from the Southern Hemisphere compared with the Northern Hemisphere?

Benefits to States

Economic Development

The NewSpace sector is fast growing and provides significant opportunity for economic growth. The global space industry was estimated at US$345b in 2016 (FAA, 2018). Even for the recently space-enabled New Zealand, the space sector contributed approximately NZ$1.69b to its economy in 2018–2019 with 5,000 direct and 12,000 indirect jobs (Deloitte, 2020).

Environmental Monitoring

The ability to place remote sensing assets in space has positive downstream effects allowing a nation to actively monitor environmental phenomena impacting industries such as agriculture (Bastiaanssen et. al., 2000) and waste management (Gao et. al., 2014).

International Relations

The space sector provides opportunities to develop international partnerships and collaborative agreements. A contemporary example of this is the Joint Letter of Intent signed between the Australian Space Agency and NASA in September 2019 to pursue exploration on the Moon and Mars (Australian Space Agency, 2019a).

Increased Independence

Currently, states such as Australia source most of its satellite data by sharing the in-space capabilities of commercial companies or international partners who own and operate the satellites (Australian Government, 2013 and Clark et. al., 2018). Reliable access to space increases the independence of States on foreign or domestic commercial and State partners.

Increased Resilience

Reliable access to space ensures that a State can quickly redeploy space assets in the event of its spacecraft or other space capability being incapacitated for example by orbital debris. Rapid reinstatement of space capability is fundamental for disaster relief operations, civil infrastructure, and military operations (Australian Government, 2016).

In considering important factors for the NewSpace industry, access to space from the Southern Hemisphere offers the following advantages when compared to the Northern Hemisphere.

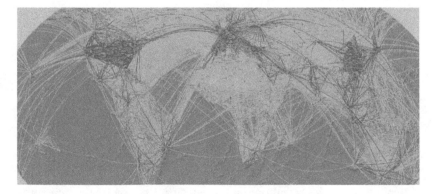

Figure 5: Air Traffic Density Worldwide in 2015 (Source: International Civil Aviation Organization, 2015).

Reduced Time to Launch

A priority for the NewSpace industry is to reduce waiting periods for customers who want to place a satellite/s into orbit. When launch occurs, airspace must be cleared to facilitate rocket passage (Davenport et. al., 2018). The density of airspace traffic in the Southern Hemisphere is significantly less than in the Northern Hemisphere (Figure 5). Launch activities from the Southern Hemisphere are therefore likely to cause less disruption to global supply chains which is important to facilitate the launch frequency that NewSpace companies desire. For example, Rocket Lab has stated it would like to launch every week by the end of 2020 (Rocket Lab, 2018).

Access to Polar Orbits

It is important to ensure that rocket stages do not fall into neighbouring nations during launch. While launches aiming for a polar orbit may be launched either North or South, launching South from selected Southern Hemisphere nations provides a much lower chance for this to occur. In the Northern Hemisphere, launching North from existing launch pads in the USA, Europe or even China will mean that the rocket will likely have to pass over the territory of other nations. In contrast, launching South from future stations in Australia, South Africa or Argentina is much safer with trajectories entirely over the ocean.

Scenario Comparison

Having identified needs and benefits of State access to space, the ways in which this may be achieved in the Southern Hemisphere can now be explored. The enabler for space access in the Southern Hemisphere is almost certain to be NewSpace companies. Despite funding of such companies being primarily private investment, strategic direction at the State level is required to support new investment and establishment. To expand on this requirement, the project team developed and discussed three different scenarios to determine one approach by which States in the Southern Hemisphere may pursue space access.

The three scenarios are:

1. Independent National Pursuit of Space Access
2. International Public-Private-Partnership (IPPP)
3. International collaborative State-based framework

Each scenario has been analyzed using the SWOT methodology. The analysis attempts to capture the ability of each scenario to enable access to space for the Southern Hemisphere as a whole. In accordance with the mission statement, political, economic, strategic and environmental factors are the key metrics. SWOT items have been identified using a dual approach consisting of an initial project team analysis followed by an expert survey to ensure all aspects were considered. For brevity, only the five items deemed most important in each category are shown. To enable a decision of the best performing scenario, a numerical evaluation using key metrics has been performed at the end of the section.

Scenario 1: Independent national pursuit of space access

In this scenario each nation in the Southern Hemisphere pursues access to space in relative isolation, supporting their NewSpace start-up companies as needed. In pursuit of their national interest, each State will either develop sovereign capability in each of the areas required for space access, or engage in international engagement for those capabilities that they decide they will not develop locally.

Strengths

- National level consensus, coordination and planning ensures efficient use of resources.
- Significant strategic benefits to States as they may have the entire space access chain in-country.
- Facilitates significant local infrastructure development.
- Harness the capabilities of other nations through commercial means.
- Builds sovereign capability in areas aligned with the national interest.

Weaknesses

- Requires significant investment both publicly and privately, meaning that smaller Southern Hemisphere states may be left behind.
- Payload capacity may not be fully utilized.
- Leads to duplication of facilities in every nation and hence subcritical utilization of facilities.
- Higher capital expenditure compared to international partnership scenarios.

Opportunities

- Invigorated Research and Development (R&D) towards technological advancements, enabling and strengthening state independance and status.
- Job growth in space-related sectors leading to increased national Gross Domestic Product (GDP).
- Inspiring the next generation of STEM students.
- Necessitates that national space legislation is updated to consider all aspects of space access.

Threats

- Market demand may not provide critical mass to ensure commercial launch business in all nations.
- Competition or confrontation between countries over issues such as launch window, space orbits, frequency allocation and ground based segments.
- Legal disputes, opposition and restriction from other state players.
- Launch sites in every nation will cause significant disruption for global shipping and air traffic.

Scenario 2: International Public - Private - Partnership (PPP)

The second scenario considered is where governments of Southern Hemisphere States and private NewSpace companies form an international entity to facilitate space access. The entity would provide launch access, ground station communication services and other services required for space access. Invested states may then implement in national policy that all governmental, academic and industrial satellites must go through this company.

Strengths

- Infrastructure will be built only where it makes the most business sense.
- Establishment of a globally recognised commercial space model.
- Sharing of infrastructure, launch-services, profits and risks.
- Consensus of operations and collaborations toward space and related R&D.
- Business decisions will drive the operation of the IPPP leading to higher efficiency access to space.

Weaknesses

- Parties involved are unable to attain respective optimal goals.
- Some member nations may miss out on space investment.
- Requires upfront investment and full negotiation from all member organisations.
- Possible conflict between public and private entities on mission. objectives, service allocations and common interests.
- Complicated profit distribution, project organization, and mechanism setup.

Opportunities

- Encourages growth of commercial space industries and affiliated business.
- Introduction of commercial capital including Private Equity (PE) and Venture Capital (VC).
- International cooperation between government, universities, and commercial entities.
- States can develop expertize in diversified areas of space activities.

Threats

- Risk of losing taxpayer money if the project fails.
- Possible conflicts between existing state and international regulations.
- Risk of creating one "super company" that completely dominates the market and reduces competitiveness.

Scenario 3: International collaborative State-based framework

The final scenario considers the development of an international framework, facilitated by a treaty, to which States of the Southern Hemisphere may become a member. The framework would aim to facilitate equitable access to space for the member states. This proposed cooperative framework would mean that each individual State need not develop sovereign capability for every aspect of space access, instead enabling states to focus on aspects suitable for their national requirements and inherent domestic skill sets. For example, one nation may be well suited to have launch capabilities, while another may focus on developing ground-based infrastructure. Each nation could provide discounted services or priority access to other member states and their organizations.

Strengths

- Offers opportunities for all States to gain space access regardless of size or economic strength.
- Facilitates economic, technological and cultural integration through international cooperation.
- Provides certainty for the NewSpace market as to which services will be required in various locations.

- Makes use of existing infrastructure and thus requires less investment on the part of single nations.
- Likely to facilitate the critical mass of satellites needed to ensure commercial viability of launch companies involved.

Weaknesses

- Small/weak economies with less investment will receive less in participation and return.
- Introduces a high level of political involvement.
- Launch states take on the majority of the risk which may cause tension.
- Will require significant effort to align permit and regulatory procedures between members.
- Not every State will have the strategic benefits that arise from sovereign launch.

Opportunities

- Opportunities for smaller countries to be involved in projects which were previously out of reach.
- Expansion of international cooperation in other fields.
- Consolidation of space industry standards.
- States may become experts in specific space areas.
- Reduction in space debris arising from consolidated payload delivery.

Threats

- Difficulty in treaty negotiation.
- Countries with launch capabilities will have dominance in cooperation and may lead to power imbalance.
- Withdrawal of member States may lead to complete stagnation or collapse of the overall arrangement.
- The regional cooperation model may be regarded as a special alliance by the international community and lead to backlash.
- Some uncertainty on how funding will work.

Evaluation

Each scenario is now evaluated for its ability to provide reliable, effective and efficient access to space for the Southern Hemisphere. The mission statement has been used to define several metrics for evaluation with respect to both the Southern Hemisphere as a whole and to individual nation states. All metrics are given equal weighting except for the metric about the ability of the scenario to facilitate space access for the entire Southern Hemisphere which has been weighted by a factor of 1.5. The project team considers this metric the most important factor. The scenarios have been assigned a score in each metric from 1 to 3, with 3 being the highest as shown in Table 5. A discussion of how the rankings were justified for each metric is contained in Appendix A.

The analysis indicates Scenario 3 best fits the evaluation criteria. That is not to say that Scenario 3 is flawless; the SWOT analysis shows there are important threats and weaknesses that exist. However, given the mission statement for the report is concerned with access to space for the Southern Hemisphere, rather than a specific State, scenario three will be discussed in detail as it is deemed to be the most appropriate by which States in the Southern Hemisphere may pursue space access.

Table 5: Evaluation of Scenarios According to Key Metrics.

	Scenarios		
Metrics	1	2	3
Facilitates reliable access to space for the entire Southern Hemisphere (Weighting 1.5x)	1	1	3
Economic Benefit	2	1	2
Environmental Impacts	1	2	2
Political Factors	2	2	2
Strategic Considerations	3	2	2
Totals	9.5	8.5	12.5

Scenario Implementation

The scenario analysis found that an international framework between member states may provide an effective way of developing reliable, effective and efficient commercial access to space for Southern Hemisphere nations. In this section the scenario is explored further through two parts. Part 1 will explore how such a framework might be implemented on a hemispheric level, while Part 2 takes a national viewpoint using Australia as a case study.

Part 1: Implementation

It is not within the scope of this report to provide a full geopolitical analysis of the implementation of the international collaborative framework from Scenario 3. Instead this section focuses on two questions:

1. Which nations in the Southern Hemisphere would likely be launch states within the framework?
2. What are the considerations surrounding the creation of the framework with respect to space legislation and the NewSpace landscape?

Geographical Location

Geographical location plays an important part in how efficient launch may be as well as determining the direction that launch could occur. For example, due to the spherical shape of the Earth, there is a significant velocity benefit gained from launching into equatorial orbits from close to the equator (Greshko, 2019). Therefore sites located close to the equator require less fuel and are more cost effective. Other geographical factors under consideration are whether there is a safe launch path East (for equatorial orbits) or South (for polar orbits). This is being determined simply by whether there are other countries in the near vicinity either East or South.

Meteorological Environment

Both low and high altitude weather effects have the ability to postpone rocket launches. In April 2019, SpaceX postponed a launch of the Falcon Heavy due to strong winds in the upper atmosphere above the launch site (Chang, 2019). Given frequent launches are a feature of the NewSpace industry, the stability of a country's meteorological environment is important. For this report, the Global Climate Risk Index (GCRI) for each nation in 2020, developed by Germanwatch, will be used as a proxy for the stability of each nation's risk of extreme events (Germanwatch, 2020).

Launch States

Notwithstanding existing definitions under international law, a launching state, in the context of the scenario, is a nation that has private launch companies operating on its territory that provide discounted or priority service to organisations or governments of other members of the framework.

It is difficult to select a subset of the most important factors that affect the suitability of a country for launch. However, for the purposes of this report we will consider the following criteria.

Infrastructure

Evaluation of suitable launching states under the international framework scenario requires consideration of launch facilities currently operating and those actively being planned. It would be advantageous to use existing facilities where possible.

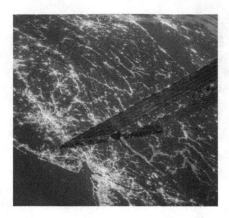

Political Stability

The political stability of a country is equally as important as the climate stability to ensure continual operation of facilities, open borders and a positive business environment. The Fragile State Index (FSI) of each nation for 2020, published by The Fund for Peace, is used as a metric for political stability (https://fragilestatesindex.org/).

Table 6 shows a summary of these metrics for a subset of eight nations that the project team has decided might be suitable to be a launching state within the framework. A country's latitude is taken as the closest point to the equator, with a latitude of zero given if the country sits on the equator. For both the CGRI and FSI, lower numbers indicate higher stability.

The nations that perform well against all criteria are New Zealand and Australia, with no countries immediately East or South of them, relatively low GCRI and with either existing or planned launch infrastructure. As a benefit, both nations are able to provide equatorial and polar launches. Argentina and Kenya are the next best placed and, together with Australia and New Zealand, form a subset of Southern Hemisphere nations that could be beneficial locations for launch infrastructure. An advantage of this set of nations is that they are spread throughout the world, mitigating risks from large travel distances and weather events closing down single sites.

Table 6: Performance of Countries Against Launch Location Criteria

Country	Latitude(°S) (Closest point to the equator - 0 if country spans equator)	Percentage ground speed change compared to equator	Safe launch path East	Safe launch path South	GCRI	FSI	Launch facilities
Argentina	22	-7.28	Yes	Yes	48.3	46	Three existing suborbital facilities (Nuclear Threat Initiative, 2020)
Australia	10	-1.52	Yes	Yes	49.5	19.7	Two future launch facilities in planning (Davis, M., 2019)
Brazil	5	-0.38	Yes	Yes	82.8	71.8	One orbital launch site (Pappalardo, 2019)
Kenya	0	0	Yes	Yes	19.6	93.5	One launch facility (Ibeh, 2019a)
New Zealand	34	-17.1	Yes	Yes	53.2	20.1	One launch facility (Rocket Lab, 2020a)
Peru	0	0	No	Yes	94		No launch facilities
South Africa	22	-7.28	Yes	Yes	53.3	71.1	One decommissioned launch facility
Indonesia	0	0	Yes	Yes	68.2	70.4	One future launch site planned (Dzulfikar, 2019)

Scenario Considerations

Consideration 1: Launch Liability

Due to Article VII of the 1967 Treaty on Principles Governing the Activities of States in the Exploration and Use of Outer Space, including the Moon and Other Celestial Bodies (Outer Space Treaty or OST), States are liable to any damage caused by launches from their territory. This means that even when the launch is conducted by private companies, the liability rests with the Nation State. In the collaborative framework concept, member nations would need to consider their respective liabilities under existing international agreements.

Consideration 2: Priority Access

One of the attractive benefits to nations in the framework is priority access for their companies to launch within the launching states of the framework. As such, member nations should ensure their national space legislation is drafted in accordance with the framework treaty, the OST and other international agreements to allow and encourage this priority access for other members.

Consideration 3: Initial Members

The international framework may not be supported by all Southern Hemisphere States from inception. Prior to framework development, a subset of nations that can form a key nucleus around which the framework can form should be considered. This would include, at minimum, one State with launch capabilities.

Consideration 4: Engagement Pathway

Scenario three considers the formation of a governing treaty at the national level. The project team recognizes that the establishment of the treaty may be politically difficult. As such it could be beneficial to begin with soft agreements at the provincial level such as the agreement between New South Wales and Luxemburg (Luxembourg Government, 2020). If successful, this can then lead to the formation of the treaty at the national level.

Consideration 5: Competition

Successful formation of new launch facilities within the framework is likely to reduce demand for launches in the Northern Hemisphere. NewSpace companies in the Northern Hemisphere may revisit launch costs to provide competition. This is significant because approximately two-thirds of payloads from 2018 and 2019 are commercial and may be freely competed on the open market (FAA, 2018 and Bryce Space and Technology, 2019 and 2020).

Part 2: Australia Case Study

Within the framework of the proposed international collaboration, Australia's likely role is to be a launching state. Part 2 of this section explores the current state of Australia's capability and discusses areas that could be addressed to better facilitate its role in the framework.

Geopolitcs

The Australian space industry is supported by its constant promotion of international cooperation, as well as the Australian Space Agency's active participation in international forums, agreements and treaties. Australia is a signatory to the five United Nations Treaties on Outer Space, and is a member of at least three space-related international forums, namely, the UN Committee on the Peaceful Use of Outer Space, the International Astronautical Federation, and the Asia-Pacific Regional Space Agency Forum (Australian Space Agency, 2019b).

By investigating Australia's partnership status internationally, it was evident that a well-established strategic space network and partnership exists. This is due to Australia's announcement of signed partnerships with other nations, with a focus area of cooperation in space. For example, the Australian Space Agency has entered into Memorandum of Understandings (MoUs) with the Centre National D'Etudes Spatiales (CNES) (Australian Space Agency, 2018a), the Canadian Space Agency (CSA) (Australian Space Agency, 2018b), the United Kingdom Space Agency (UKSA) (Australian Space Agency, 2018c), the United Arab Emirates Space Agency (UAESA)(Australian Space Agency, 2019c)], and the Italian Space Agency (Australian Space Agency, 2019d).

In terms of Australia's strategic international partnership with Southern Hemisphere countries, the New Zealand Ministry of Business, Innovation and Employment entered an agreement with the Australian Space Agency on 22 of October 2019. The agreement provides a framework for collaborative activities in areas of common interest, in addition to facilitating the exchange of information, technology and personnel (The New Zealand Ministry of Business, Innovation and Employment, 2019).

For geopolitical factors to be strengthened and facilitated between Australia and Southern Hemisphere countries, considering Australia as a launch country, the following should be taken into account: Undertake and pursue MoUs with Southern Hemisphere programs, starting with immediate neighbours and moving outwards. Australia has signed an MoU with one Southern Hemisphere country, namely New Zealand; however, it needs to further develop and execute a plan for space cooperation agreements in regards to launch activities, including manufacturing and transportation of launch vehicles, in providing logistical support for space stations and other space-related projects with Southern Hemisphere countries.

The need for fuel transport and processing agreements to avoid the increased risk for supply reliability and security, as Australia is already dependent on imports to meet the growth in demand for transport fuels, where shipments to Australia has increased to offer supply flexibility and reliability (Australian Institute of Petroleum, n.d.). It is recommended that this also includes assessing the viability of integration with Asian markets, as it has the lowest cost source of alternative fuel supply and has excess supply capacity.

Partnership agreements/framework for payloads providing complete access to equatorial launch sites (in the case of Australia) or polar (in the case of South-East Asia), in a manner that is consistent with Australia Space (Launches and Returns) Act 2018.

Policy & Legislation

Currently, the Australian Space Agency is responsible for regulating Australian space and high power rocket activities, and to facilitate international arrangements affecting space regulation. To ensure alignment with innovative technology advances, Australia amended its Space Activities Act 1998, the recent being known as the Space Activities (Launches and Returns) Act 2018 which entered into force on 31 August 2019 (Australian Space Agency, 2020).

The renewed Act includes provisions focusing on launch activities, namely, the approval requirement to authorize any space activity or launch of high power rockets in Australia, and the need for any Australian nationals to have their space activities approved if conducted outside Australia. It also provides provisions in terms of "liability for damage caused by space objects or high power rockets", the registration of space objects in its national register, and the investigation procedures in case of accidents or incidents. The investigation procedure is, however, only concerned with (a) space objects launched from a launch facility in Australia or from an aircraft that is in the airspace over Australian territory; (b) a space object returned to a space or area in Australia; or (c) a high power rocket launched from a facility or place in Australia (Australian Government, 2019).

As a member state of the collaborative framework, Australia would need to consider the following:

The Australian Space Agency, when defining missions, would strongly consider giving preference for launch vehicles or other existing space transport systems of member states. It would also make their launch facilities available at a pre-agreed cost clarified in the collaborative framework, so as to share the benefits of a partnership.

Australia's domestic legislation would need to be updated to expand its context within the Australian territory to cover association activities related to further space development, including the manufacturing, transportation, testing, integration, ground stations, placing in orbit, and control of satellites, as well as other launch related activities.

The Australian Space Agency would continue to provide leadership at a national level, while also providing support to private companies and research and development institutions. Specifically, a growth enabler would be the development of a strategic policy that sets direction, international presence, and coordination of scientific launch activities (Clark et. al., 2018).

Exchange of legal expertise in the development of the convention draft and the overall establishment of the framework would need to be considered, to ensure alignment with national laws and regulations in addition to Australia's international commitments.

The Australian Space Agency could ensure the accident and incident investigation process to include optional exchange of expertise from other member states in the collaboration, in the case where a failure of launch requires or would benefit from additional insight.

Research Landscape

For an international collaborative State-based framework to be successful, there will need to be a strong Research and Development (R&D) focus within the Southern Hemisphere. This focus is particularly important for the NewSpace sector, which is driven by new innovation. It is not within the scope of this study to assess the current research landscape of the whole southern hemisphere; Instead, this section focuses only on the research landscape of Australia, and how it could benefit the International collaborative State-based framework.

Government

The Australian Government actively supports R&D, both for the prosperity of the country generally (Innovation and Science Australia, 2017) and specifically for the space sector (Office of Minister Karen Andrews, 2019). The Australian Government established the Australian Space Agency (ASA) in July 2018 with the intention of supporting the growth of Australia's space industry and the use of space across the broader economy (Australian Space Agency, 2018d); the ASA has specified to grow the industry from AU$3.9b to AU$12b by 2030 (Australian Space Agency, 2019e).

The Australian Government has provided AU$600m in funding to date to specifically grow the Australian space sector (Office of Minister Karen Andrews, 2019). In relation to an International collaborative State-based framework, increased innovation and R&D within Australia would be positive to the whole network of countries. It is expected that the growth of the Australian space sector would trickle out to neighbouring countries through industry interconnections.

The ASA has programs to support relevant research into products and services for the space sector; this includes a AU$15m international investment initiative (Australian Space Agency, 2019) and a AU$19.5m space infrastructure fund (Australian Space Agency, 2019). This funding should allow growth of the NewSpace industry within Australia, and enable further international collaborations. Notably, the ASA has pledged AU$150m to support business and researchers to join and deliver capabilities for NASA's Moon to Mars activities (Office of Minister Karen Andrews, 2019). In the context of the international collaborative framework, the ASA could consider future funding to be pledged specifically for Southern Hemisphere space relationships, if partnerships were so developed.

Academia & research

Academic research is strong in Australia with 40 Australian universities who engage in research activities (Universities Australia, n.d.). It is not within the scope of this study to cover the engagement of these universities with the space sector. One notable domestic research resource is the Australian Government established SmartSat Cooperative Research Centre (CRC), which is backed by AU$245m in public and private cash and in-kind contributions (CIO, 2019). The SmartSat CRC provides a platform for collaborative industry-led research between research organisations, business, and government harnessing Australia's space capabilities in advanced satellite technologies and applications (SmartSat CRC, n.d.) . While this CRC requires domestic partners, international companies and research organisations can be involved in research projects under certain conditions (Australian Government, 2020). This approach provides an opportunity for expanding research projects between the network of countries in the International collaborative State-based framework. Otherwise, under the implementation of the collaborative framework, it would be expected many opportunities for collaborative research projects could be organized independently between research institutions within the Southern Hemisphere.

Industry

A space technology business cluster is geographically established in South Australia, with at least 60 organisations as of 2016 (Space Industry Centre, 2016). In the context of NewSpace, a large portion of these companies are start-ups, referred to collectively as a start-up ecosystem. The ecosystem allows companies to leverage interconnections, strengthen networks and support technology transfer. One example includes the Delta-V NewSpace Alliance which promotes partnerships between established and emerging Australian NewSpace companies (Delta-V, n.d.).

The existence of these partnerships demonstrates that the Australian Space Industry is able to cooperate in a future Southern Hemisphere system of collaboration between multiple companies.

Australia has an established culture of innovation, shown through government initiatives and academic programs supporting R&D. The recent growth of the space sector (Australian Space Agency, 2019f), particularly in South Australia where a space technology ecosystem of start-ups has developed, is an example of success driven by innovation within the country. The implementation of the international collaborative framework would be well received by the space technology ecosystem, thereby benefiting other interested and participating parties.

Infrastructure

At the most basic level, the launch process is as depicted in figure 6. As a launch state, Australia will need to possess civil infrastructure and capabilities to facilitate the steps within the red box. Here, we will compare the current capabilities within Australia with the expected needs.

Manufacturing capabilities

There is currently limited capability within Australia to manufacture launch vehicles. Gilmour Space Technologies, a NewSpace startup company, is one example of an Australian company planning to develop launch vehicles in Australia. In 2017, Gilmour Space Technologies collected AU$24m of funding to develop small rockets for suborbital and LEO flights, using low-cost hybrid rockets for small satellite launch (Crunchbase, 2020 and ACIL Allen Consulting, 2017).

Australia has potential to increase its manufacturing presence for launch vehicles, especially in the case of NewSpace, where launch vehicles are often much smaller than traditional designs. New rockets being made to facilitate cheaper access to space for small satellites are often manufactured with advanced manufacturing techniques such as metal 3D printing. As an example, Rocket Lab in New Zealand uses 3D printing to manufacture the engine in its Electron launch vehicle (Rocket Lab, 2020b). Australia currently has a high level of capability on advanced technologies such as additive manufacturing and machine learning (ACIL Allen Consulting, 2017), meaning that Australia is well placed to advance this capability.

Given that civil launch facilities in Australia are yet to operate, a significant opportunity exists to integrate the manufacturing capabilities directly into the launch sites. Both of Australia's proposed launch sites are not integrated complexes such as Rocket Lab, given that the owners of both future sites will facilitate multiple launch vehicles (Government of South Australia, 2018 and Bishton, 2018). As a result, the facilities will require access to a large number of parts for different launch vehicles. With the increasing prevalence of 3D printed parts increases in NewSpace launch vehicles a business model could arise whereby the rocket parts are printed at the launch facility and assembled, therefore reducing transport costs and wait times.

Launch facilities

Australia currently has four existing or planned launch facilities, including:

1. Woomera Instrumented Range, South Australia - Air Force Base
2. Whalers Way Orbital Launch Complex, South Australia
3. Koonibba Test Range, South Australia
4. Arnhem Space Center, Northern Territory

Koonibba Test Range, owned by Southern Launch, is not an orbital launch site. Rather, it will be used as a test range facility to help universities, companies and agencies to have a more secure launch and to recover their payloads and rockets (Lee, 2020). Instead, Southern Launch's Whalers Way Orbital Launch Complex will allow rocket manufacturers to launch their own satellites into polar orbits., with the payload capacity of launch vehicles placed at Southern

Launch's between 50 and 400 kg (Dillon, 2018). ELA's Arnhem Space Center has entered into discussions with NASA to launch sounding rockets in order to test instruments that will be mounted on satellites and spacecrafts; however, it is planned that it will eventually support orbital-class launch vehicles (Waters, 2019).

Woomera Instrumented Range was Australia's original site used for orbital launches; since capability was abandoned, it has been used for sounding rockets, as well as testing and evaluation of various Defence assets (Australian Government). Two of these facilities, Arnhem Space Center and Whalers Way Orbital Launch Complex are of interest in this report as these are planned to support dedicated Smallsat launch vehicles (ADM, 2019 and Spence, 2019)

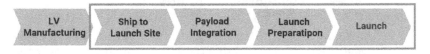

Figure 6: Overview of Launch Process

It remains to be proven whether market demand can sustain one polar and one equatorial launch site in Australia. Regardless, each launch site should be equipped with all infrastructure required to facilitate new space launches. Infrastructure to be considered at launch facilities includes:

Fuel storage: Both solid fuel and liquid fuel rocket engines are used in new launch vehicles being developed (FAA, 2018). This means that the launch facilities will need to have cryogenic storage chambers to store liquid fuels, as well as procedures and methods to store solid rocket motors.

Hardware storage: The launch facility at Whalers Way, for example, is aiming to provide a range of launch pads to facilitate multiple types of launch vehicles (Damp, 22 January). Furthermore, NewSpace launch businesses are aiming to launch very frequently, up to weekly in the case of Rocket Lab (Rocket Lab, 2018). In order to ensure a regular cadence of launch at the Australian launch facilities, a storage facility large enough to hold enough stock (both launch vehicles and satellite payloads) for several months worth of launches is required.

Transportation

As the majority of launch vehicles are made in the Northern Hemisphere, shipping of launch vehicles to Australia is likely to take a long time. Implementing an aircraft runway may alleviate this issue by allowing rapid airfreight transfer directly to the launch facility. There also may be potential in creating or investigating landing runways for some spacecraft, given the United Nation's interest in finding a Landing Site for their upcoming Orbital Missions (United Nations).

The consideration of transportation and storage of explosives or dangerous goods for the site journey to and on site, must comply with the "Australian Code for the Transport of Explosives by road and rail" (NTC Australia, 2018) as well as state-level legislation. Some considerations recommended in the code are the vehicle design, container type, portable tanks, storage limits, dangerous goods compatibility, emergency procedures, markings, insurance, route to be driven and the control of potential igniters such as an isolation switch for the transport vehicles batteries. These requirements would be essential minimum considerations for rocket system transportation into and around Australia, and thus it must be considered.

unsplashed.com

Testing facilities

Test and Evaluation (T&E) capabilities within Australia will be beneficial to its role as a launch state. Extensive design of T&E regimes for components and mission systems is critical to ensure component or system integrity, safety and useability (MITRE, 2020).

The two main types of testing involved in T&E are Non-destructive Testing (NDT) and Destructive or Dynamic Testing (Sampson, 2018). NDT methodologies such as Ultrasonic Testing (UT) and Acoustic Emission Testing (AE) are designed to identify flaws in components without changing their state (TWI, 2020). In contrast, destructive or dynamic testing methods such as vibration and impact testing are used to identify the limits of materials or components[s] (NTS 2020). Often, NDT methods are used to complement each other to verify results (ENG 2018).

The Department of Defence White Paper 2016 (Australian Government, 2016) acknowledges underinvestment in current testing facilities. Australia has several testing facilities which are mostly within the university and Defence sectors. Currently, only one testing facility, Defence's Proof and Experimental Establishment site at Port Wakefield, environmentally tests explosives (propellants) – a capability that is required to test rocket motors (Pavic and Jezierski, 2017). Given that this facility is likely unavailable to the commercial civilian market, either the site would need to be expanded and allowed to be more accessible to industry, or a new civilian site created.

Two of the emerging trends that NewSpace has facilitated are reusability of rocket stages, as pioneered by SpaceX (SpaceX, 2020), and advanced manufacturing as pioneered by Rocket Lab (Rocket Lab, 2020). With the assumption that future launch vehicles will continue with this trend, NDT facilities will be crucial to qualify recovered rocket stages for reuse. Positioning NDT infrastructure in the near vicinity of the launch facilities is optimal to facilitate the rapid launch cadence of NewSpace. The T&E infrastructure needs of the planned Whalers Way launch facility in South Australia and the Arnhem Space Center in the Northern Territory should be a priority for operational safety and integrity of rocket motors, satellites, launch gantry systems, and site specific asset requirements. The project team encourages nations to consider on-site NDT testing infrastructure for new space facilities.

Integration facilities

As the NewSpace market matures it will be heavily driven by smaller size satellites with cheaper and faster access to space (NASA, 2020). As such, the integration facility needs to accommodate corresponding scales of operation, to integrate satellites with the launch vehicle. Integration facilities for rockets and satellites include clean rooms, fueling stations and storage rooms. Having a large number of integration facilities near the launch facility will allow the launch facility to handle more launches at a time and reduce the time-to-orbit for customers.

Ground stations

The Australian Government currently has ground station network infrastructure that covers Deep Space Exploration, Meteorology, Astronomy and Earth Observation related activities. For example, the advanced capability in Astronomy is showcased through the Square Kilometre Array Project in Western Australia (ACIL Allen Consulting, 2017). The private sector also contributes to space activities by providing ground stations for communications, earth observations, GNSS services and space situational awareness. Australia has pre-existing satellite operation capabilities, showcased by the establishment of satellite laser ranging and tracking station at Mount Stromlo. This facility is used for ranging and tracking operating satellites and space debris in space (ACIL Allen Consulting, 2017). There are at least 26 active ground stations (OPTUS, 2020 and Nova Systems, 2019) in Australia that are used to communicate with and track satellites.

Australia's advantages on minimal light pollution, radio noise interference and well-developed ground network infrastructures is critical to provide real-time temporal resolution for any ground station (CSIRO, 2018). These advantages have opened up many collaborations with major space agencies, including the European Space Agency's New Norcia Deep Space Antenna that is located near Perth and NASA's Canberra Deep Space Communication Complex (CSIRO, 2018). As such, it would be beneficial to leverage these advantages for other countries, if a partnership were created.

Report Limitations

The analysis of the International collaborative State-based framework scenario explores an innovative, alternative approach for Access to Space, not currently implemented in the Southern Hemisphere. However, due to the time and resource limitations placed on the project, the team acknowledges that a fully comprehensive analysis of the topics covered could not be presented. Some of the limitations the team has identified include:

Legal Analysis

There will be significant legal challenges and considerations in the establishment of an international treaty framework such as that presented in scenario 3. This report has briefly touched on the consideration of State liability, however a much more comprehensive analysis of legal implications would be necessary. Further research would be required into specific facets of policies and regulations, for example how the framework would factor into International Traffics in Arms Regulations (ITAR).

Member States

The Scenario Implementation analysis is made on the assumption that the International collaborative State-based framework will be accepted by sufficient nations within the Southern Hemisphere to realize the benefits of the scenario. That is, without a large enough number of countries signed onto the framework, then the benefits of the framework will be minimal and may not outweigh the difficulties of organising the framework. Furthermore, no analysis was done to determine what number of countries, and which countries with key capability, would be necessary to sign the framework for it to be beneficial for all parties involved, compared to the sovereign access to space approach.

Launch State Determination

This report analysed suitable launch states in the southern hemisphere, however the range of analysis parameters was quite limited. If a study was commissioned to answer the question of suitable launch states, it should look at specific sites rather than a country as a whole.

Roles of States in the Framework

This report has only explored considerations for launching states in the proposed collaborative framework. The role of members of the framework who do not pursue launch capability needs to be explored in detail before any comprehensive evaluation of the scenario could be completed.

Conclusion & Recommendations

This report has explored potential methods in which the Southern Hemisphere may continue to develop access to space. In doing so, the project team has taken the viewpoint that space access for as many nations as possible is more desirable than access for a single State. Evaluation as shown that the implementation of an international State-based framework may offer the best opportunity to facilitate this access when considering strategic, economic, political and environmental factors.

The role that NewSpace will play in pursuing this access is significant; the market survey indicates that the Smallsat market is increasing in size across the space business sector and will continue to do so. A case study of Australia has identified considerations for nations looking to encourage NewSpace launch companies, from infrastructure requirements to potential policy changes. From the case study, the team has concluded that Australia has potential to develop launch capabilities and would be well suited as a launching state within the framework.

The project team acknowledges the limitations of the report analysis scope, however, believes that all aspects of the mission statement have been addressed within the constraints of the project.

Based on the research conducted in this Team Project, the project team recommends the following:

1. Further studies on the implementation of an international collaborative State-based framework should be completed addressing the limitations set out in the previous section.
2. International engagements and treaty negotiations to facilitate such a collaborative framework should consider the liability and risk taken on by launch states.

3. Launch facilities built to service the NewSpace market should have easy access to NDT infrastructure to facilitate emerging reusable launch vehicles.

4. Nations seeking to develop their space access capabilities should ensure their industry, government and research sectors support a strong R&D environment.

5. In the event of implementing a similar international framework to the one discussed in the report, States with similar launch potential as Australia should consider taking on the role of a launch state.

References

Avio, 2020. Avio for the Southern Hemisphere Space Studies Program. [presentation] (personal communication 28 January 2020)

ACIL Allen Consulting, 2017. Australian Space Industry Capability - A Review. [pdf] Available at: <https://www.industry.gov.au/sites/default/files/2019-03/australian_space_industry_capability_-_a_review.pdf> [Accessed 7 February 2020].

Australian Government, 2013. Australia's Satellite Utilisation Policy. [pdf] Canberra: Commonwealth of Australia. Available at: <https.//www.industry.gov.au/sites/default/files/May%202018/document/pdf/australias_satellite_utilisation_policy.pdf> [Accessed 3 February 2020].

Australian Government, 2016. Defence White Paper 2016. [pdf] Canberra: Commonwealth of Australia. Available at: <https://www.defence.gov.au/WhitePaper/Docs/2016-Defence-White-Paper.pdf> [Accessed 6 February 2020].

Australian Government, 2019. Space (Launches and Returns) Act 2018. [online] Available at: <https://www.legislation.gov.au/Details/C2019C00246> [Accessed 8 February 2020].

Australian Government, 2020. Funding for short-term, industry-led research collaborations. [online] Available at: <https://www.business.gov.au/Grants-and-Programs/Cooperative-Research-Centres-Projects-CRCP-Grants> [Accessed 7 February 2020].

Australian Government, n.d. About the Woomera Prohibited Area. [online] Available at: <https://www.defence.gov.au/woomera/> [Accessed 7 February 2020].

Australian Institute of Petroleum, n.d. Facts about the Australian transport fuels market. [pdf] Available at: <https://aip.com.au/sites/default/files/download-files/2017-09/Facts%20about%20the%20Australian%20Transport%20Fuels%20Market_2015_1.pdf> [Accessed 8 February 2020].

Australian Space Agency, 2018a. Memorandum of Understanding between the Australian Space Agency and The Centre National D'Etudes Spatiales for Strategic Cooperation on Space. [pdf] Available at: <www.industry. gov.au/sites/default/files/2019-04/mou-australian-space-agency-and-centre-national-detudes-spatiales.pdf> [Accessed 8 February 2020].

Australian Space Agency, 2018b. Memorandum of Understanding Between Australian Space Agency and Canadian Space Agency Regarding Cooperation in the Exploration and Use of Space for Peaceful Purposes. [pdf] Available at: <www.industry.gov.au/sites/default/files/2019-04/mou-australian-space-agency-and-canadian-space-agency.pdf> [Accessed 8 February 2020].

Australian Space Agency, 2018c. Memorandum of Understanding Between Australian Space Agency and United Kingdom Space Agency Regarding Civil Space Cooperation for Peaceful Purposes. [pdf] Available at: <www. industry.gov.au/sites/default/files/2019-04/mou-australian-space-agency-and-united-kingdom-space-agency.pdf> [Accessed 8 February 2020].

Australian Space Agency, 2018d. Australian Space Agency Charter. [pdf] Available at: <https://www.industry.gov.au/sites/default/files/2018-10/australian-space-agency-charter.pdf> [Accessed 7 February 2020].

Australian Space Agency, 2019a. Australia to support NASA's plan to return to the Moon and on to Mars. [online] 22 September. Available at: <https://www.industry.gov.au/news-media/australian-space-agency-news/australia-to-support-nasas-plan-to-return-to-the-moon-and-on-to-mars> [Accessed 6 February 2020].

Australian Space Agency, 2019b. International collaboration on space. [online] Available at: <www.industry.gov.au/strategies-for-the-future/australian-space-agency/international-collaboration-on-space> [accessed 8 February 2020].

Australian Space Agency, 2019c. Memorandum of Understanding Between Australian Space Agency and the United Arab Emirates Space Agency Regarding Space Cooperation for Peaceful Purposes. [pdf] Available at: <www.industry.gov.au/sites/default/files/2019-04/mou-australian-space-agency-and-united-arab-emirates-space-agency.pdf> [Accessed 8 February 2020].

Australian Space Agency, 2019d. Memorandum of Understanding Between Australian Space Agency and the Italian Space Agency Regarding Space Cooperation for Peaceful Purposes. [pdf] Available at: <www.industry. gov.au/sites/default/files/2019-11/mou-australian-space-agency-and-agenzia-spaziale-italiana.pdf> [Accessed 8 February 2020].

Australian Space Agency, 2019e. Growing the space industry through cutting-edge downstream capabilities. [pdf] Available at: <https://www.industry.gov.au/news-media/australian-space-agency-news/growing-the-space-industry-through-cutting-edge-downstream-capabilities> [Accessed 7 February 2020].

Australian Space Agency, 2019f. Australian Civil Space Strategy 2019-2028. [online] Available at: <https://www.industry.gov.au/data-and-publications/australian-civil-space-strategy-2019-2028> [Accessed 7 February 2020].

Australian Space Agency, 2019g. International Space Investment initiative. [online] Available at: <https://www.industry.gov.au/data-and-publications/international-space-investment-initiative> [Accessed 7 February 2020].

Australian Space Agency, 2019h. Space Infrastructure Fund. [online] Available at: <https://www.industry.gov.au/data-and-publications/space-infrastructure-fund> [Accessed 7 February 2020].

Australian Space Agency, 2020. Regulating Australian space activities [online] Available at: <www.industry.gov.au/regulations-and-standards/regulating-australian-space-activities> [Accessed 8 February 2020]

Bastiaanssen, W., Molden, D.J., Makin, I.W., 2000. Remote sensing for irrigated agriculture: examples from research and possible applications, Agricultural Water Management, 46(2), pp.137-155.

BBC News, 2018. European Space Agency teams with ICEYE Finnish start-up. BBC News, [online] 26 March. Available at: <https://www.bbc.com/news/science-environment-43544211> [Accessed 9 February 2020]

Sampson, B., 2018. Introduction to non-destructive testing. Aerospace Testing International [online] Available at: <https://www.aerospacetestinginternational.com/features/introduction-to-non-destructive-testing.html> [Accessed 7 February 2020]

Boadle, A., 2018. Brazil space station open for small satellite business. Reuters. [online] Available at: <https://www.reuters.com/article/us-space-brazil-usa/brazil-space-station-open-for-small-satellite-business-idUSKCN1LV007> [Accessed 11 February 2020].

Bryce Space and Technology, 2019. Smallsats by the numbers: 2019. [pdf] Available at: <https://brycetech.com/downloads/Bryce_Smallsats_2019.pdf> [Accessed 10 February 2020]

Bryce Space and Technology, 2020. Smallsats by the numbers: 2020. [pdf] Available at: <https://brycetech.com/downloads/Bryce_Smallsats_2020.pdf> [Accessed 10 February 2020]

Waters, C., 2019. 'World First': Startup wins NASA deal to launch rockets from Australia.The Sydney Morning Herald, [online] 31 May. Available at: <https://www.smh.com.au/business/small-business/world-first-startup-wins-nasa-deal-to-launch-rockets-from-australia-20190531-p51t8g.html> [Accessed 7 February 2020]

Chang, K., 2019. Falcon Heavy Launch Postponed by SpaceX. The New York Times [online] April 10. Available at: <https://www.nytimes.com/2019/04/10/science/falcon-heavy-launch-spacex.html> [Accessed 08 February 2020]

CIO Australia, 2019. UniSA secures $245M for smart satellite centre. [online] Available at: <https://www.cio.com/article/3498611/unisa-secures-245m-for-smart-satellite-centre.html> [Accessed 7 February 2020].

Henry, C., 2019. SpaceX submits paperwork for 30,000 more Starlink satellites. SpaceNews, [online] 15 October. Available at: <https://spacenews.com/spacex-submits-paperwork-for-30000-more-starlink-satellites/> [Accessed 9 February 2020]

Clark et al., 2018. Review of Australia's Space Industry Capability: Report from the Export the Reference Group for the Review. Australian Government. [online] Available at: <https://www.industry.gov.au/data-and-publications/review-of-australias-space-industry-capability> accessed 8 February 2020

Crunchbase, 2020. [database] Available at: <https://www.crunchbase.com/> [Accessed 7 February 2020]

CSIRO, 2018. A roadmap for unlocking future growth opportunities for Australia: 2018. [pdf] Available at: <https://www.csiro.au/~/media/Do-Business/Files/Futures/18-00349_SER-FUT_SpaceRoadmap_WEB_180904.pdf> [Accessed 9 February 2020]

Daniel Bishton, 2018. Australia's new space race. Spatial Source, [online] Available at: <https://www.spatialsource.com.au/gis-data/australias-new-space-race> [Accessed 7 February 2020]

Damp, L., Southern Launch Small Rockets big future, [presentation] (personal communication 22 January 2020)

Datta, A., 2017. The NewSpace Revolution: The emerging commercial space industry and new technologies. Geospatial World, [online] 8 January. Available at: <https://www.geospatialworld.net/article/emerging-commercial-space-industry-new-technologies/> [Accessed 05 February 2020].

Davenport, C., Muyskens, J., Shin, Y. and Ulmanu, M. 2018. Gridlock in the Sky. The Washington Post [online] 12 December. Available at: <https://www.washingtonpost.com/graphics/2018/business/spacex-falcon-heavy-launch-faa-air-traffic/> [Accessed 07 February 2020]

Davis, M., 2019. Northern launch site could transform Australia's role in space. The Strategist, [online] 20 Dec. Available at <https://www. aspistrategist.org.au/northern-launch-site-could-transform-australias-role-in-space/> [Accessed 09 February 2020]

Deloitte, 2019. New Zealand Space Economy: Its value, scope and structure. Deloitte, [online] <https://www.mbie.govt.nz/dmsdocument/7307-new-zealand-space-economy-its-value-scope-and-structure> [Accessed 07 February 2020]

Delta-V. NEWSPACE ALLIANCE. [online] Available at: <http://www. deltavspacehub.com/#space20> [Accessed 7 February 2020]

DelPozzo, S., Williams, C. and Doncaster, B., 2018. 2019 Nano/Microsatellite Market Forecast, 9th Edition. [pdf] Available at: <https://www.spaceworks. aero/wp-content/uploads/Nano-Microsatellite-Market-Forecast-9th-Edition-2019.pdf> [Accessed 08 February 2020]

Department for Trade and Investment, 2020. Southern Launch satellite facility in South Australia to be the first of its kind in Australia. [online] Available at: <https://dti.sa.gov.au/news/southern-launch-satellite-facility-in-south-australia-to-be-the-first-of-its-kin> [Accessed 05 February 2020].

Dzulfikar, L.T., 2019. Indonesia's first spaceport in Biak, Papua, set to become first equatorial launching site in the Pacific. The Conversation. [online] Available at: <http://theconversation.com/indonesias-first-spaceport-in-biak-papua-set-to-become-first-equatorial-launching-site-in-the-pacific-127499> [Accessed 08 February 2020]

Eckstein, D., Kunzel, V., Schafer, L., Winges, M., 2019. Global Climate Risk Index 2020. Kalserstr:Germanwatch. Available at <https://germanwatch. org/sites/germanwatch.org/files/20-2-01e%20Global%20Climate%20 Risk%20Index%202020_14.pdf> [Accessed 08 February 2020]

ELIE, ENG 2018. Types of Non-Destructive Testing. Nucleom. [online] Available at: <https://nucleom.ca/en/knowledge/types-of-non-destructive-testing/> [Accessed 7 February 2020]

Escher, A., 2018. Inside Planet Labs' new satellite manufacturing site. Tech Crunch, [online] 15 September. Available at: <https://techcrunch. com/2018/09/14/inside-planet-labs-new-satellite-manufacturing-site/> [Accessed 9 February 2020]

Etherington, D., 2019. SpaceX will now offer dedicated 'rideshare' launches for small satellites. Tech Crunch [online]. Available at: <https:// techcrunch.com/2019/08/05/spacex-will-now-offer-dedicated-rideshare-launches-for-small-satellites/> [Accessed 05 February 2020]

Federal Aviation Administration (FAA), 2018. The Annual Compendium of Commercial Space Transportation: 2018. Bryce Space and Technology. [online] Available at <https://www.faa.gov/about/office_org/headquarters_offices/ast/media/2018_ast_compendium.pdf> [Accessed 7 February 2020]

Foust, J., 2019. Rocket Lab to build second launch pad in New Zealand. [online] Available at: <https://spacenews.com/rocket-lab-to-build-second-launch-pad-in-new-zealand/> [Accessed 10 February 2020]

Gilmour Space, 2019. Scaling up for the next generation of rocket technology. [online] Available at: <http://www.gspacetech.com/post/scaling-up-for-the-next-generation-of-rocket-technology> [Accessed 7 February 2020]

Government of South Australia, 2018. Southern Launch satellite facility in South Australia to be the first of its kind in Australia. [online] Available at: <https://dti.sa.gov.au/news/southern-launch-satellite-facility-in-south-australia-to-be-the-first-of-its-kin> [Accessed 7 February 2020]

Greshko, M., 2019. Rockets and rocket launches explained. National Geographic [online] January 4. Available at <https://www.nationalgeographic.com/science/space/reference/rockets-and-rocket-launches-explained/> [Accessed 08 February 2020]

Gunters Space Page, 2020. Launch Sites. [online] Available at: <https://space.skyrocket.de/directories/launchsites.htm> [Accessed 06 February 2020]

Guo, Y., Feng, N., Christopher, S.A., Kang, P., Zhan, B. and Hong, S., 2014. Satellite remote sensing of fine particulate matter (PM2.5) air quality over Beijing using MODIS, International Journal of Remote Sensing, 35(17), pp. 6522-6544

Ibeh, J., 2019a. Kenya and Italy Close In On Signing Ownership Deal In Respect Of Luigi Broglio Space Centre. Space in Africa [online] 19 July. Available at: <https://africanews.space/kenya-and-italy-signing-ownership-deal-luigi-broglio-space-centre/> [Accessed 11 February 2020]

Ibeh, J., 2019b. SANSA Receives USD 7 Million To Revamp Old Space Facilities. Space in Africa [online] 2 July. Available at: <https://africanews.space/sansa-receives-usd-7-million-to-revamp-old-space-facilities/> [Accessed 11 February 2020]

Innovation and Science Australia, 2017. Australia 2030 Prosperity through Innovation. [pdf] Available at: <https://www.industry.gov.au/sites/default/files/May%202018/document/extra/australia-2030-prosperity-through-innovation-summary.pdf?acsf_files_redirect> [Accessed 7 February 2020].

International Civil Aviation Organisation (ICAO), 2015. ICAO Gallery. [online] Available at: <https://gis.icao.int/gallery/graphics/FLOWCHART2015.JPG> [Accessed 08 February 2020]

Lee, S., 2020. Rocket range to test suborbital launches over outback South Australia for space research. [online] Available at: <https://www.abc.net.au/news/2020-01-31/rocket-range-to-test-suborbital-launches-over-outback-sa/11915570> [Accessed 7 February 2020]

Louis Dillon, 2018. Australia's first commercial launch facility finds home. [online] Available at: <https://www.spaceconnectonline.com.au/r-d/3113-australia-s-first-commercial-launch-facility-finds-home> [Accessed 7 February 2020]

Luxembourg Government, 2020. New South Wales and Luxembourg sign MoU on future space activities. [Press Release] 3 February. Available at: <https://gouvernement.lu/fr/actualites/toutes_actualites.gouvernement%2Ben%2Bactualites%2Btoutes_actualites%2Bcommuniques%2B2020%2B02-fevrier%2B03-newsouthwales-mou.html> [Accessed 11 February 2020]

Masunaga, S., 2018. A new wave of satellites in orbit: Cheap and tiny, with short lifespans. [online] Available at: <https://phys.org/news/2018-08-satellites-orbit-cheap-tiny-short.html> [Accessed 10 February 2020]

Moon, M., 2019. Virgin Orbit will launch satellites from Guam. [online] Available at: <https://www.engadget.com/2019/04/11/virgin-orbit-launch-site-guam/> [Accessed 10 February 2020]

Myriota, 2019. Myriota Partners with Tyvak to Develop and Launch Next Generation Nanosatellites [press release] 7 February. Available at: <https://myriota.com/2019/02/07/myriota-partners-with-tyvak-to-develop-and-launch-next-generation-nanosatellites/> [Accessed 9 February 2020]

Nanosats Database, 2020. Nanosatellite constellations [online] Available at: <https://www.nanosats.eu/img/fig/Nanosats_constellations_2020-01-06.pdf> [Accessed 9 February 2020]

NASA, 2020. New Space: the "Emerging" Commercial Space Industry. [pdf] Available at: <https://ntrs.nasa.gov/archive/nasa/casi.ntrs.nasa.gov/20140011156.pdf> [Accessed 7 February 2020]

Nova Systems, 2019. Nova IGS network delivers space ground connectivity to Peterborough, SA. [online] Available at: <https://novasystems.com/news/nova-igs-network-delivers-space-ground-connectivity-to-peterborough-sa/> [Accessed 7 February 2020]

NTC Australia, 2018. Australian Code for the transport of dangerous goods by road & rail. [pdf] Available at: <https://www.ntc.gov.au/sites/default/files/assets/files/Australian-Code-for-the-Transport-of-Dangerous-Goods-by-Road%26Rail-7.6.pdf> [Accessed 7 February 2020]

NTS, 2020. Dynamic Testing. [online] Available at: <https://www.nts.com/services/testing/dynamics/> [Accessed 7 February 2020]

Nuclear Threat Initiative, 2020. Argentina: Facilities. [online] Available at: <https://www.nti.org/learn/countries/argentina/facilities/> [Accessed 10 February 2020]

Office of Minister Karen Andrews, 2019. Space agreements boost international collaboration. [Online] Available at: <https://www.minister.industry.gov.au/ministers/karenandrews/media-releases/space-agreements-boost-international-collaboration> [Accessed 7 February 2020].

OPTUS, 2020. Optus Earth Stations. [online] Available at: <https://www.optus.com.au/about/network/satellite/earth-stations> [Accessed 7 February 2020]

Pappalardo, J., 2019. Can the U.S. Save Alcantara, Brazil's Cursed Spaceport?. Popular Mechanics [online] Available at: <https://www.popularmechanics.com/space/rockets/a26884062/us-brazils-cursed-spaceport-alcantara/> [Accessed 10 February 2020]

Rocket Lab, 2018. Rocket Lab enters high frequency launch operations. [online] Available at: <https://www.rocketlabusa.com/news/updates/rocket-lab-enters-high-frequency-launch-operations/> [Accessed 06 February 2020]

Rocket Lab, 2020a. Our Launch Facilities. [online] Available at: <https://www.rocketlabusa.com/launch/launch-sites/> [Accessed 08 February 2020]

Rocket Lab, 2020b. Rocket Lab: The small firm that launched the 3D-printed space revolution. [online] Available at: <https://www.technologyreview.com/s/613792/rocket-lab-the-small-firm-that-launched-the-3d-printed-space-revolution/> [Accessed 7 February 2020]

Seradata, 2020. SpaceTrak. [database] Available at: <https://www.seradata.com/products/spacetrak/> Accessed [07 February 2020]

SmartSat CRC. Supporting Australia's Space Industry. [pdf] Available at: <https://www.deakin.edu.au/__data/assets/pdf_file/0020/533126/World-leading-research-in-Australias-universities.pdf> [Accessed 7 February 2020]

South Australian Space Industry Centre, 2016. Space Innovation and Growth Strategy (South Australia) Action Plan 2016-2018 [pdf] Available at: <https://www.sasic.sa.gov.au/docs/default-source/default-document-library/sasic-action-plan.pdf> [Accessed 7 February 2020].

SpaceDaily, 2018. India seeks to reduce satellite launch cost. Space Daily, [online] 24 January. Available at: <https://www.spacedaily.com/reports/Price_War_in_Space_India_Adopts_Technologies_to_Reduce_Satellite_Launch_Cost_999.html> [Accessed 11 February 2020]

SpaceX, 2020. Reusability. [online] Available at: <https://www.spacex.com/reusability-key-making-human-life-multi-planetary> [Accessed 7 February 2020]

Spence, A., 2019. Australian rocket launch site given major project status. The Lead [online], 19 September. Available at: <http://theleadsouthaustralia.com.au/industries/space/australian-rocket-launch-site-given-major-project-status/> [Accessed 10 February 2020]

Strahinja, S.P and Jezierski, P., 2017. Transition of the JPEU engineering management system towards the explosive ordnance safety program enhances safety of the explosives ordnance test and evaluation. [pdf] Available at: <https://www.unsw.adfa.edu.au/conferences/sites/conferences/files/uploads/0900%20Pavic%20Parari%202017%20-%20Pavic_0.pdf> [Accessed 7 February 2020]

The Jakarta Post, 2019. Indonesia to build the nation's first spaceport in Papua.The Jakarta Post [online], 12 November. Available at: <https://www.thejakartapost.com/news/2019/11/12/indonesia-to-build-the-nations-first-spaceport-in-papua.html> [Accessed 10 February 2020]

TWI, 2020. What is non-destructive testing (NDT)? - Methods and Definition. [online] Available at: <https://www.twi-global.com/technical-knowledge/faqs/what-is-non-destructive-testing> [Accessed 7 February 2020]

United Nations, 2019. Orbital Space Mission. [online] Available at: <https://www.unoosa.org/documents/doc/psa/hsti/CFI_SNC_LandingSite_final.docx> [Accessed 11 February 2020]

Universities Australia, n.d. World-leading research in Australia's universities. [pdf] Available at: <https://www.deakin.edu.au/__data/assets/pdf_file/0020/533126/World-leading-research-in-Australias-universities.pdf> [Accessed 7 February 2020]

Appendix A
Evaluation Discussion for Scenario Comparison

Metric 1: Facilitates reliable access to space for the entire Southern Hemisphere (1.5x)

A key weakness of Scenario 1 is that smaller and less powerful Southern Hemisphere nations are unlikely to have the resources to develop reliable access to space in isolation. The establishment of a Public Private Partnership is slightly better than Scenario 1 in this regard, however, it still requires significant upfront investment from members which may lead to exclusion of some countries. In comparison, Scenario 3 has strong potential to enable this access for all member states.

Metric 2: Economic Benefit

Scenario 1 requires the greatest individual investment by States to achieve the full launch capability, however, has great potential to deliver economic returns to the nation as infrastructure is built in-country. Scenario 2 is weak in this regard as the IPPP may decide it is more efficient not to have infrastructure built in a particular nation's territory, and reducing the economic return to the nation. Scenario 3 is somewhere in the middle.

Metric 3: Political Factors

None of the scenarios are particularly good or bad in regards to political opportunities versus difficulties. Scenario 3 presents the greatest political opportunity for states, however, it also presents the most challenges. Scenario 1 does not provide quite as much opportunity for international diplomacy although as space access is pursued by individual nations, there are far fewer political impediments. Scenario 2 is also given a score of two because while it facilitates some international engagement, there will be political difficulty in establishing the partnership which all has to be negotiated upfront.

Metric 4: Environmental Impacts

Scenario 1 has the worst environmental footprint of the three scenarios as there will be significant duplication of infrastructure in every State. Scenarios 2 and 3 are both much more efficient in this regard and will limit impact on the natural environment. In our market analysis we have seen that the rate of growth of the Smallsat industry is projected to level off in the near future. There is a significant risk that market demand will not be high enough to service launch sites in many countries leading to empty space on rockets and more space debris per satellite launched.

Metric 5: Strategic Considerations

Scenario 1 carries a clear advantage in this metric as States will have complete control over how they align their access to space with their national interest. Scenario 3 is the worst performer as the international framework means that Nations should be guided by what is best for the entire hemisphere.

Anu Rajendran
Brianna Ganly
Joshua Kahn
Rebekah Holliday
Samuel Hobbs
Steven Despotellis
Zvonko Vidos

Hongxing Zuo
Kaimin Zhou
Na Zhang
Qiwei Guo
Shideng Yang
Shiyin Lyu
Si Li
Xiangyu (Sean) Li
Yiming Zhou
Yonggang Luo

Christian Thaler-Wolski

Guy Marcus

Preeti
Sujith Rajan

Andrea Perin
Luca Santachiara

Ibnu Rusydi

Sasiluck Thammasit

Fatima AlShamsi

CPSIA information can be obtained
at www.ICGtesting.com
Printed in the USA
BVHW030304160321
602630BV00003B/18